Bob Artley's Seasons on the FARM

Bob Artley's Seasons on the FARM

From the Newspaper Series "Memories of a Former Kid"

By Bob Artley
Foreword by Paul Gruchow

Voyageur Press

Copyright © 1981, 2003 by Bob Artley

All rights reserved. No part of this work may be reproduced or used in any form by any means—graphic, electronic, or mechanical, including photocopying, recording, taping, or any information storage and retrieval system—without written permission of the publisher.

Cover Design by JoDee Mittlestadt
Printed in China

03 04 05 06 07   5 4 3 2 1

Library of Congress Cataloging-in-Publication Data available
Artley, Bob.
  [Cartoons]
  Bob Artley's seasons on the farm / by Bob Artley ; foreword by Paul Gruchow.— Reprinted ed.
     p. cm.
  "From the Newspaper Series "Memories of a Former Kid"."
  Originally published by the Iowa State Univ. Press in 1981 as Cartoons.
     ISBN 0-89658-609-X (pbk. : alk.paper)
  1. Artley, Bob. 2. Cartoonists—Iowa—Biography. 3. Farm life—Iowa—Caricatures and cartoons. 4. American wit and humor, Pictorial. I. Title: Seasons on the farm. II. Title.
  NC1429.A76A2 2003
     741.5'973—dc21
                          2003008397

To Ginny

Distributed in Canada by Raincoast Books, 9050 Shaughnessy Street, Vancouver, B.C. V6P 6E5

Published by Voyageur Press, Inc.
123 North Second Street, P.O. Box 338, Stillwater, MN 55082 U.S.A.
651-430-2210, fax 651-430-2211
books@voyageurpress.com
www.voyageurpress.com

Educators, fundraisers, premium and gift buyers, publicists, and marketing managers: Looking for creative products and new sales ideas? Voyageur Press books are available at special discounts when purchased in quantities, and special editions can be created to your specifications. For details contact the marketing department at 800-888-9653.

Contents

Foreword   7

Spring   9

Summer   51

Autumn   103

Winter   145

Foreword

These drawings speak to us out of the universal province of memory.

Some of what we read strikes us as pretty but artificial; we enjoy it and forget it. Some of what we read strikes us as incredible or bizarre; these things we easily forget, too. Some of what we encounter in books or newspapers seems simply beyond judgment—irrelevant, or unremarkable, or outside our experience—and that we pass by.

But when the fully realized artist speaks to us, he speaks directly to our own memories. When we read such an artist, we remember thinking the same thought once, or experiencing once an identical moment, or having once just such an emotion. We are reminded by these shocks of recognition of what Joseph Conrad called the "solidarity of mankind." We are reminded of the tie that binds all humans together—the dead to the living, and the living to the unborn. This is why art is our great weapon against loneliness, which is the beginning of despair.

This may seem too much baggage to attach to Bob Artley's "cartoons," but it isn't. We may think we are charmed and moved by them because they are nostalgic, because they hark back to a quieter and lovelier time, because they are about Iowa, or about farm boys, or something. It is only incidentally so.

These drawings are, in fact, much better than that. At their core, they are not at all about long-ago farm life. They are about growing up, one of the half-dozen great themes of art, a theme Artley has addressed in a fresh and beautiful way. As his title suggests, he has discovered the child buried within himself, and he has dared to share that child truthfully with the rest of us. He reminds us of our own child-ness. He has put us into joyful touch with that forgotten part of ourselves.

For this, we are in Artley's immense and long-lasting debt.

Paul Gruchow

SPRING

SUMMER

AUTUMN

WINTER